The Art of Life

ABSTRACT

This work links Egyptian Philosophy, Physics and Quaternion Mathematics. The Egyptians saw science as the study of God's Works, a spiritual activity bringing one closer to the Divine. The Egyptian Civilization was built on Righteousness - Right Actions and Right Angles! A Grand Unification Theory is introduced based upon a Life quaternion and a quaternion Spacetime change operator.

1. Egyptian Spacetime and Zodiac Ages [1]

Space

The Egyptians realized the Earth was a Rotating Sphere. They divided the Circumference of the sphere into 360 degrees and each degree they divided into 60 Minutes of arc of the Earth's Circumference. Each Arc-Minute of Circumference, they called a Mile. The Arc-Minute/Mile was measured to be 6000 feet or 4000 Cubits by observing the rotation of the Stars. The Arc-Minute was further divided into 60 Arc-Seconds; thus an Arc-Second was measured to be 100 Feet.

Time

The Egyptians invented a measure for Time called the day. Night originally defined the start of the day ("and the evening and the day was the first day"). Star movement was the original clock. The day was first divided into two halves Night and Day, then star clocks divided the night into 12 hours by tracking star risings and the day and night became each 12 hours for the 24 hour day. As measurement technology improved, each hour was broken into 60 minutes of time and each minute of time was broken into 60 Seconds.

Faster than the Speed of sound!

With a measure for distance, the Mile, and a measure for time, the hour, the Egyptians observed that the Earth rotated 900 Miles per Hour at the Equator; faster than the 720 Miles per Hour speed of sound in air. Thus the Egyptians could calculate the

1.1 Size of the Earth: 24 Hours x 900 Miles/Hour = 21,600 Miles Circumference!

Age of Aquarius

The Egyptians by observing the Stars discovered the precession of the Equinox. The precession of the Equinox causes the coming of spring to move backwards in the Zodiac (Star Belt) one sign about every 2160 years. This change they called the " a new Age or EON." The Egyptians recorded the Ages by the Zodiac, The Age of the BULL (Taurus), was followed by the Age of the RAM or LAMB (Aries), which was followed by the Age of the FISH (Pisces) during the Roman Empire. These Ages were marked by the use of the Bull, Lamb and Fish as religious symbols.

2. Life

To the Ancient Egyptians, Life was the foundation of the Universe. The Universe Lives and is immortal. The Egyptians recognized "space-time" and defined two types of change; change in time "ter" and change in space "del". The word "delta" is Egyptian for "far (del) land (ta). Our modern use of "tel" such as in television and telegraph derives from the Egyptian "del" meaning far, ("d" and "t" are linguistic substitutes as are "r and l " and "f and v", but this would take us too far from our topic). Work and Force are Changes to the essence, Life. Change is the fundamental Law of the Universe. Life, Work, Force and Change are the key concepts in Egyptian science. Modern science has not yet recognized the wisdom of the Egyptians and their realization that the Universe is about Life and is Alive.

The closest modern science has come to a concept of Life is the concept of 'action". Action came to prominence when William Rowan Hamilton (1800's) demonstrated that physical laws could be expressed as minimizing or maximizing the Action of a system. Action has dimensions of work-time. Action is a "time-based" concept. In Quantum Theory, Planck introduced the action concept to modern physics, in his constant of action "h ". The concept of action is close to the concept of Life and there are those who say that all physical activity including "life", involves the physics of action. While I agree with this view, the units of action involve time, while the Universe is spatial. However, one can define a physical spatial concept of Life $L = ch$, with units of work-feet, by multiplying action by the speed of light c. Life can be shown to be the central concept in the Universe and Physics.

Spacetime points $P = (ct) + (Ix + Jy + Kz)$, are quaternion. . The scalar part is "ct", the product of the speed of light (c) and time (t). William Rowan Hamilton developed quaternions in 1843 as a means of rotating a line in three-space. He found out that it takes four dimensions, one scalar unit and three vectors to solve this problem. Life Theory is based on quaternions. Life is also a quaternion, the sum of one scalar Ls and three vectors $Lv = ILx + JLy + KLz$. Quaternions coefficients (Ls, Lx, Ly and Lz) can be complex scalars. Mathematically, the complex numbers are a subset of quaternions.

2.1 Life $L = Ls + ILx + JLy + KLz = Ls + Lv$

Some Life Constants
2.2 Life: $Lo = hc = 4\mu$ evf = 4 microelectron-volt-feet = work x feet.
2.3 WaveTemp: $WT = Lo/k = 46.5$ milli feet K, 46.5 milli feet-degree K

Important Life Temperatures and Wavelengths
2.4 Water Freezing 0C = 273K: Wavelength = 46.5m fK/273K = 170u feet
2.5 Human Body Temp 98.6F = 310K: Wavelength = 46.5m fK/310K = 150u feet
2.6 Water Boiling 100C = 373K: Wavelength = 46.5m fK/373K = 125u feet

3. Change happens

Change was a Divine Concept to the Egyptians and was called Xeper. Change has two aspects, Time and Space. Time is a scalar and Space is a vector. To accommodate this I defined a quaternion Change Operator X, called Xeper.

3.1 Xeper: $X = T + \nabla = d/cdt + Id/dx + Jd/dy + Kd/dz = d/cdt + \nabla$

William Rowan Hamilton invented del ∇, the vector change operator.

.

A short diversion on Quaternion may be helpful:
Hamilton's Rules for Quaternions:
3.2 $I^2 = J^2 = K^2 = IJK = -1$
3.3 $IJ = - JI, JK = - KJ, KI = - IK$
3.4 Conjugate of L: $L' = Ls - Lv$
3.5 Norm of L: $LL' = L'L = Ls^2 + Lx^2 + Ly^2 + Lz^2$
3.6 Inverse of L: $L^{-1} = L'/$Norm of L, such that $LL^{-1} = 1$

Vector Change Definitions
3.7 $\nabla.Lv$ (∇ dot Lv) is the Divergence of vector Lv
3.8 ∇Ls is the Gradient of scalar Ls
3.9 $\nabla x Lv$ (∇ cross Lv) is the Curl of vector Lv

Vector Associativity Identities
3.10 $0 = \nabla.\nabla x Lv$:Curls are transverse to change.
3.11 $0 = \nabla x \nabla Ls$:Gradients are longitudinal to change.
3.12 $\nabla(\nabla.Lv) = \nabla^2 Lv + \nabla x \nabla x Lv$:Gradient of Divergence $= \nabla^2 Lv + \nabla x \nabla x Lv$

4. WORK is the Change of Life:
4.1 $W = X L = (T + \nabla) L = (TLs - \nabla.Lv) + (TLv + \nabla Ls + \nabla x Lv)$
Substituting $T = d/cdt = F/c$ and $L = hc$, gives:
$W = X L = (dhs/dt - \nabla.Lv) + (dhv/dt + \nabla Ls + \nabla x Lv)$
$W = X L = (hsF - \nabla.Lv) + (hvF + \nabla Ls + \nabla x Lv)$

Conservation of Life is given by $0 = X L$:
4.2 $0 = X L = (TLs - \nabla.Lv) + (TLv + \nabla Ls + \nabla x Lv)$

For Quaternions to be zero, both the scalar and vector part must each be zero.
4.2.1 $0 = dhs/dt - \nabla.Lv$ (scalar dependency)
4.2.2 $0 = dhv/dt + \nabla Ls + \nabla x Lv$ (vector dependency)

5. FORCE is the Change of Work, the Curvature of Life, and The Art of Life!
In a way Force is the most fundamental concept because Force is what animates the Universe. No Force, No Curvature. In general force is Hyperbolic!

5.1 Force $F = X^2 L = ((T^2 - \nabla^2)Ls - 2T\nabla.Lv)) + ((T^2 - \nabla^2)Lv + 2T(\nabla Ls + \nabla xLv))$

Equilibrium Condition or zero Force:

5.2 $0 = X^2 L = ((T^2 - \nabla^2)Ls - 2T\nabla.Lv) + ((T^2 - \nabla^2)Lv + 2T(\nabla Ls + \nabla xLv))$

Force is a Quaternion wave equation consisting of a scalar wave
$((T^2 - \nabla^2)Ls - 2T\nabla.Lv)$ and a vector wave $((T^2 -\nabla^2)Lv + 2T(\nabla Ls + \nabla xLv))$.
The scalar wave is a longitudinal $(2T\nabla.Lv)$, meaning the vibration of the wave is in
the direction of the wave. Sound is an example of a scalar wave. The vector wave is a
transverse (perpendicular) wave $(2T\nabla xLv)$, meaning the vibration is transverse to
the direction of the wave. Water ripples from a rock dropped in a pool are an
example of transverse waves as is Light and Radio waves.

The Force under the Conservation Condition, $0 = X L$, is Elliptical!
5.3 $F = - (T^2 + \nabla^2) L$ at $0=X L$.
5.4 $F <0$: Maximum, A Peak or Cap
5.5 $F >0$: Minimum, A Pit or Cup
5.6 $F =0$: Minimax = Saddle Point

The transport (motion) part of the wave is the differential "interval":

5.7 $(T^2 - \nabla^2)(Ls + Lv) = (d^2/(cdt)^2 - (d^2/dx^2 + d^2/dy^2 + d^2/dz^2))(Ls + Lv)$

The Gradient is responsible for the refraction of the wave due to scalar fields :

5.8 $2T\nabla Ls$ (∇Ls = Gradient of Ls)

6. "And then there was LIGHT!"
Examining the Force Equation 5.1 provides an "explanation" of the wave-particle
duality of light. Light is a Quaternion consisting of a scalar longitudinal wave
(denoted by $\nabla.Lv$) and a vector transverse wave (denoted by ∇xLv). The
photoelectric effect is a manifestation of the longitudinal wave and diffraction is a
manifestation of the "wave effect". Both effects are waves representing Force and
the curvature of Life.
James Clerk Maxwell presented Maxwell's Equations and brilliantly predicted that
light was electromagnetism! For over 100 years, Maxwell's Equations have defined
modern physics and introduced asymmetry that has constrained physicists thinking.
Electromagnetism is symmetric as would be suspected but for the respect for
Maxwell's Equations. The correct and complete Equations of Electromagnetism are
symmetric and consist of a single Conservation Condition, $0 = XB$!

ElectroMagnetism
E

E Electric Field Strength
e electric permittivity
D=eE Electric Flux Density
E=cB

H

H Magnetic Field Strength
u magnetic permeability
B=uH Magnetic Flux Density
H=cD

Electromagnetic Relations
$c^2 = 1/ue$
$cu = 1/ce = z = 1/y =$ impedance:

$zH = cuH = cB = E$
$yE = ceE = cD = H$

6.1 $0 = X\ B = (dBs/cdt - \nabla.Bv) + (dBv/cdt + \nabla Bs + \nabla xBv)$
Multiplying by c gives (and substituting E= cB)
6.2 $0 = XE = (dBs/dt - \nabla.Ev) + (dBv/dt + \nabla Es + \nabla xEv)$
Multiplying by the dielectric e gives (eE = D):
6.3 $0 = X\ D = (dDs/cdt - \nabla.Dv) + (dDv/cdt + \nabla Ds + \nabla xDv)$
Multiplying by c gives (and substituting H = cD)
6.4 $0 = XH = (dDs/dt - \nabla.Hv) + (dDv/dt + \nabla Hs + \nabla xHv)$

Maxwell's Equations differ from the above by assuming dBs/cdt to be zero, thus
ignoring the scalar ("Charge/Matter") field and have the sign of xHv reversed;
Maxwell's Equations are a representation of the experimental state-of-the-art of
1900 century experimental physics. The vector mathematics, Maxwell used was
created by Hamilton as a part of Hamilton's Quaternions. Quaternions I believe are
an even more brilliant achievement, providing the mathematical microscope into
Life and the operation of the Universe. J. Willard Gibbs's Vectors displaced
Hamilton's Quaternions from physics in 1900. Will quaternions rise again?

The Ether
Maxwell believed that light passed traveled through an ether in passing from the
source to the destination. The Ether is the result of 2 new constants;
Mo = 500 atto volt seconds and Qo= 1.326 atto coulombs. The product of Mo and
Qo is Planck's Constant h = 663 atto atto joule seconds. The ratio of Mo and Qo is
the "free space" impedance zo = 377 Ohms.

7. Life, Light and Electromagnetic Wave Characteristics

Band	FREQUENCY # waves/second	DENSITY # waves/foot	Application	Work
GG/s	X/s	G/f	X-RAYS	Kev
MG/s	P/s	M/f	Ultraviolet	eV
KG/s	T/s	K/f	Heat & Light	mev
G/s	G/s	1/f	RADAR	uev
mG/s	M/s	m/f	RADIO	nev
uG/s	K/s	u/f	INFRA RADIO	pev
nG/s	1/s	n/f	Earth Waves	fev

RADAR	# waves/second	# waves/foot
Ka	27-40 G/s	27-40/f
K	18-27 G/s	18-27/f
Ku	12-18 G/s	12-18/f
X	8-12 G/s	8-12/f
C	4-8 G/s	4-8/f
S	2-4 G/s	2-4/f
L	1-2 G/s	1-2/f

8. Life and Light, Constants and Measures

8.1 Egyptian mile = 6000 feet = 4000 cubits

8.2 knot = Egyptian mile/hour

8.3 Light speed c = 1 Giga feet/sec =10 Mega miles/minute = 600 Mega miles/hour

8.4 Earth Rotation speed = 1500 feet/second = 15 miles/minute = 900 miles/hour

8.5 Earth Circumference = 900 miles/hour x 24 hours = 21,600 miles

8.6 Earth Radius = 21,600/2pi = 3438 miles

8.7 Planck's Constant h = 663 atto atto joule seconds = 4 femto evs

8.8 Boltzman's Constant k = 86 uev/K

8.9 Life Quantum Lo = hc = 4 micro electron-volt -feet = .48 Pico Pico # square feet

8.10 WaveTemp Quantum WT = Lo/k = 46.5 m fK

8.11 Gravitational Constant G = 1 atto square miles/# or
Universal Pressure P = 1/G = 1 X #/square miles

8.12 Joule = 6.25 X eV = cal/4 = kal/4K = 3/4 # feet =1/8K # miles

9.0 EGYPTIAN ARITHMETIC [2]

The Egyptians developed Binary Arithmetic, avoiding the need to remember long multiplication tables (9.1). They also used Inverse numbers that allowed them to use multiplication rather than division for problems. For example sharing 3 cakes with 5 persons or 3 divided by 5 or 3/5, the Egyptians used Inverse Numbers (9.2).

9.1 Binary Arithmetic 5 x 13

```
x        13

1/       13

2        26

4/       52
---------------
5        65
```

9.2 DIVISION: 3/5 =?   Or 5x ? = 3

```
 x               5

/2              2 /2

/10              /2
----------------
(/2 /10)  =    3/5
```

Division : 19/7

```
1             7

2+            14

/2+           3 /2

/7+           1

/14+          /2
----------------------
2 /2 /7 /14 =   19 /7
```

9.3 My formula for determining Inverse doublets: /n = /(n+1) /n(n+1)

/1 = /2   /2

/2 = /3   /6

/3 = /4 /12

/4 = /5   /20 = /6 /12

/5 = /6   /30

/6 = /7 /42 = /8   /24 = /9 /18

/7 = /8 /56

/8 = /9   /72   = /10   /40 = /12   /24

/9 = /10 /90 = /12 /36

/10 = /11   /110 = /12   /60 = /15   /30

10. Egyptian Handy Measures [3]

Binary Fractions, The Foot, The Pint and The Pound

The Egyptians used a binary counting system called the:

10.1 Horus-Eye Fractions: 1/2, 1/4, 1/8, 1/16, 1/32 and 1/64.

WEIGHT: Pint = Pound of Water
The Egyptians subdivided the Mile into 6000 feet and used the foot as a basic measure. A foot cubed was their basic unit of VOLUME, the Artaba. The weight of a cubic foot of WATER is 64 pounds (#), the WEIGHT of a Talent. A Talent was said to be half the load a man could carry, two Talents balanced on a two cubit (3 foot) bar. 1/64th of a cubic foot is a Pint volume and weighs one pound (#); The Pint Volume and Pound Weight are the HANDY Measures used by the Egyptians.
VOLUME
Bushel = foot cubed = 64 Pints = 8 Gallons = Artaba
Gallon = foot/2 cubed = 8 Pints = Artaba/8 = Egyptian Hekat measure for grain
Pint = foot/4 cubed = Palm Cubed = Egyptian Beer Jug = Pound = Hekat/8
Pint/2 = Cup = 8 oz = Hekat/16
Pint/4 = Gill = 4 oz = Cup/2 = Hekat/32
Pint/8 = 2 oz = Cup/4 = whiskey shot glass = Hekat/64

LENGTH
Earth's Circumference = 360 Degrees x 60 Minutes x 1 Mile = 21,600 Miles
Mile = 6000 feet
Stadia = 600 feet
Second of Earth's Circumference = 100 feet
Fathom = 6 feet
Cubit = 1 1/2 feet = 6 Palms
Foot = 2/3 Cubit = 4 Palms = 16 Fingers = 256 parts
Palm = foot/4 = 3 inches = 4 Fingers = 64 parts
AREA
1 square mile = 36 million square feet = 625 acres
1 Acre = 57,600 square feet = 240x240 square feet
Quarter Acre = 14,400 square feet = 120x120 square feet
40 Acres is approximately a quarter mile squared (39.0625 Acres)

References

[1. Lockyer, J. Norman THE DAWN OF ASTRONOMY. Montana: Kessinger Publishing Co., P.O. Box 160, Kila, MT 59920. 1992]

[2. Gillings, Richard MATHEMATICS IN THE TIME OF THE PHARAOHS. New York: Dover Publications, Inc. 31 East 2nd Street, Mineola, NY 11501, 1972.]

[3. Tompkins, Peter SECRETS OF THE GREAT PYRAMID. New York: Harper & Row, Publishers inc., 10 East 53rd Street, New York, NY 10022, 1971.]

Recommended Reading List:

- 1. Massey, Gerald. ANCIENT EGYPT THE LIGHT OF THE WORLD vol 1& 2. Kila, MT:
- Kessinger Publishing Co., 1992 (First published 1907).
-
- 2. MACKENZIE, DONALD A. MYTHS OF CHINA AND JAPAN. NEW YOR, NY:
- RANDOM HOUSE, 1994.
-
- 3. Budge, E.A. Wallis. EGYPTIAN LANGUAGE. NEW YORK NY:
- DOVER PUBLICATIONS, 1983 (First published 1910).
-
- 4. Budge. E. A. Wallis. THE EGYPTIAN BOOK OF THE DEAD. NEW YORK NY:
- DOVER PUBLICATIONS, 1967 (First published 1895).
-
- 5. KARENGA, MAULANA. THE BOOK OF COMING FORTH BY DAY. Los Angeles CA:
- University of Sankore Press, 1990.
-
- 6. MILTON, KWASI. The Genesis of Writing HIEROGLYPHICS. Hampton VA:
- United Brothers Graphics and Printing Co, 1992.
-
- 7. Lockyer, J. Norman. The Dawn of Astronomy. Kila MT:
- Kessinger Publishing Co, 1992.
-
- 8. Paine, Thomas. The Life and Major Writings of Thomas Paine (Ed. Phillip S. Foner).
- New York NY: Carol Publishing Group, 1993.
-
- 10, Smith, Adam. The Wealth of Nations. New York NY: Random House
- (The Modern Library), 1967.
-
- 11. TZU, SUN. THE ART OF WAR (Translated by Samuel B. Griffith). New York NY:
- OXFORD UNIVERSITY PRESS, 1963.
-
- 12. HERODOTUS. THE HISTORIES. NEW YORK NY: PENGUIN BOOKS, 1972,
-

- 13. ERMAN, ADOLPH. LIFE IN ANCIENT EGYPT. NEW YORK NY: DOVER PUBLICATIONS, 1971 (First published 1894)
- 14. NOSTRADAMUS. THE PROPHECIES OF NOSTRADAMUS (EDITED BY ERIKA CHEETHAM). NEW YORK NY: BERKLEY BOOKS, 1983.
-
- 15. NOSTRADAMUS. PREDICTS THE END OF THE WORLD (EDITED BY RENE NOORBERGEN. NEW YORK NY: PINNACLE BOOKS, 1983.
-
- 16. GILLINGS, RICHARD J. MATHEMATICS IN THE TIME OF THE PHAROAHS.
- NEW YORK NY: DOVER PUBLICATIONS, 1982.
-
- 17. SERTIMA, IVAN VAN. EGYPT REVISTED: JOURNAL OF AFRICAN CIVILIZATIONS VOL 10 SUMMERS 1989. NEW BRUNSWICK NJ: TRANSACTION PUBLISHERS, 1991.
-
- 18. WHITE, J.E. MANCHIP. ANCIENT EGYPT ITS CULTURE AND HISTORY. NEW YORK NY: DOVER PUBLICATIONS, 1970.
-
- 19. TOMPKINS, PETER. SECRETS OF THE GREAT PYRAMID. NEW YORK NY:
- HARPER & ROW, 1978.
-
- 20, DIOP, CHEIKH ANTA. CIVILIZATION OR BARBASRISM. NEW YORK NY:
- LAWRENCE HILL BOOKS, 1991.
-
- 21. KARENGA, MAULANA. SELECTIONS FROM THE HUSIA: SACRED WISDOM OF ANCIENT EGYPT. LOS ANGELES CA: THE UNIVERSITY OF SANKORE PRESS, 1989.
-
- 22. SERTIMA, IVAN VAN. AFRICAN PRESENCE IN EARLY AMERICA. NEW BRUNSWICK NY: TRANSACTION BOOKS, 1992.
-
- 23. BURNS, EDWARD MCNALL. WESTERN CIVILIZATIONS VOL 1 (ED ROBERT LERNER & STANDISH MEACHAM. NEW YORK NY: W.W. NORTON, 1980.
-
- 24. BURNS, EDWARD MCNALL. WESTERN CIVILIZATIONS VOL 2 (ED ROBERT LERNER & STANDISH MEACHAM. NEW YORK NY: W.W. NORTON, 1980.
-
- 25. ALDRED, CYRIL. THE EGYPTIANS. LONDON ENGLAND: THAMES & HUDSON LTD, 1988.

Quaternion Physics

ABSTRACT:
A four-dimensional Action-related variable called "Life" and a Spacetime Differential Operator provide the keys to unifying Quantum and General Relativity Theory. Electromagnetism is the first derivative of Life. Maxwell's Equations are derived from the same Differential Operator and a quaternion Electric Field. Planck's Constant, h and the "Free space" Impedance, zo, define Maxwell's Ether. The Ether flux "m" is 500 atto voltseconds and the Ether Charge "e" is 1.326 atto Coulombs. This results from h= me and zo=m/e

1.Numbers in 4 dimensional Spacetime

Numbers in physics must handle two kinds of relationships, scalar, which give size information and vectors, which give direction information in space. The combination of a scalar and a vector give a four dimensional space. This combination is called a quaternion (Hamilton [1]). Quaternions are ideally suited to physics, providing a 4-dimensional complex division algebra. More information on quaternions is given in the Mathematical Appendix.

2. Dimensional Units of Spacetime

Spacetime units are "spatial", for example "feet". Dimensional Analysis thus demands that the "time" portion of Spacetime also be in "feet". The scalar constant "c", the speed of light, converts time units (t) to space units, (Minkowski [2]). For example a point in Spacetime is given as $p = ct + Ix + Jy + Kz$, a quaternion with the scalar $ps = ct$ and vector direction $pv = Ix + Jy + Kz$. This point p can be simplified to $p = ps + pv$, where ps is the scalar part and pv the vector part. The square of the point p is also a quaternion:

$$(1) \quad p^2 = (\, (ct)^2 - (x^2 + y^2 + z^2)\,) + 2ct(Ix + Jy + Kz) = (ct)^2 (\, (1 - (v/c)^2\,) + 2v/c)$$

The scalar part of p^2 is called the "interval". The interval describes the surface of the point. For a positive surface, the interval is greater than zero. This implies that c, the speed of light is the limiting speed and the defines the Lorentz transformation. When the surface vanishes at the limiting speed, the square of the point is a pure vector.

Quaternions preserve closure. The square of p is the rotation of p about itself around the direction pv.

3. Change Operators in Spacetime, $X = T + \nabla$

Capturing Change in Spacetime led me to develop a Spacetime Change Operator, $X = T + \nabla$, by adding the scalar change, $T = d/cdt$ to the traditional vector change $DEL = \nabla = Id/dx + Jd/dy + Kd/dz$. X is the sum of a scalar and a vector and thus a quaternion. X provides a 4 dimensional Change operator for 4 dimensional Spacetime. Hamilton developed the vector Change Operator ∇, but did not add to it the Time dimension Change Operator $T = d/cdt$. Even today, in Physics, the time Change does not incorporate the c in cdt. This means change has units of feet and units of seconds as opposed to just feet.

(2) $X = T + \nabla = d/cdt + Id/dx + Jd/dy + Kd/dz$

X is the first derivative of Change, and is the sum of the scalar and vector change. The second derivative of Change is Curvature X^2:

(3) $X^2 = (T^2 - \nabla^2) + 2T\nabla$

Curvature, X^2, shows that the wave equations are intrinsic to the nature of change in quaternion Spacetime, and is independent of the variable undergoing change. The scalar of Curvature is seen to be what physicists call the D'Alembertian function, or the transport function in wave equations. The third derivative of Change, Jerk, is also a quaternion:

(4) $X^3 = (T^2 - 3\nabla^2)T + (3T^2 - \nabla^2)\nabla$

4. Quantum and Relativity Theory in Spacetime
A. Quantum Theory

Action , h, is the fundamental concept in Quantum Theory. However, h, is a "time unit" concept, with dimensions energy-seconds. Given that the units of Spacetime are "space units" like feet, the action Spacetime complement is called Life L=ch. L and h are quaternions related by the speed of light c. "c" relates many concepts in Spacetime: $c = E/B = H/D = z/u = y/e = L/h$. The numerator variables are the "space units" and the denominator variables are the complementary "time units", which are transformed by c to the space units. This very simple velocity ratio (c) between Electromagnetic units and Electrostatic units, led Maxwell to conclude that "Light was Electricity!" Maxwell also thought that light carried energy and that the energy must be "somewhere" between the time it left the radiating body and arrived at the radiated body. He therefore believed the energy was "somewhere" in the intervening medium, thus the "Ether". And Maxwell, believed that the Ether must exist as the transmission medium for light. The search for the Ether, gave rise to much new physics, including Einstein's Theory of Special Relativity. Einstein concluded that the Ether is superfluous. He did not address Maxwell's reasoning.

Maxwell's Equations for radiating energy did not predict the pattern of radiation given by actual materials. To "explain" the actual experiments, Planck introduced the "quantum hypothesis" that radiation energy was proportional to the frequency, hf and the proportion was Planck's constant h! This was the founding of Quantum Theory, a theory of energy or work. Planck's hypothesis, had no theoretical foundation justifying the experimental deviations from Maxwell's Theory. What I propose here is a theoretical foundation for Planck's Quantum hypothesis. This theory is simply that Maxwell's Ether is the scalar field in a quaternion Spacetime Physics. Scalars are "orphans" of physics. They are there, but no one knows their parents. Hamilton invented Quaternions while attempting to rotate a vector in three space. Vectors were the objects of physics, invented by Hamilton and used by Maxwell. Hamilton introduced the scalar to accomplish the goal of rotating a vector in three space. This was made manifest in his famous formula $I^2 = J^2 = K^2 = -1$. The square of the unit vector is a scalar! This scalar was essential and thus quaternions consist of a scalar and three vectors. The scalar makes quaternions closed, in that the product of a quaternion is a quaternion, unlike vectors where the product of vectors can be a non-vector, a scalar. This mathematical nicety called Closure has been overlooked in mathematics and physics for years. Today, physicists still give too little respect to the scalar, even though scalars are forcing their way into Quantum Field Theory. The theory presented here is simply that physics is about quaternions consisting of a scalar and three vectors. Planck's hypothesis is about energy. Maxwell's light is also about energy. How are they related? My answer is that the fundamental variable of physics is a variable I call Life! Life is a quaternion consisting of a scalar Ls and vector Lv. Life has dimensions energy-feet and is related to h, Action by the speed of light c, such that Life L= ch, where h is also a quaternion.

(5) L = Ls + Lv = c(hs + hv)

With Life L and the Change Operator X, Quantum Theory is seen to be the first derivative of Life:

(6) work = XL = (dLs/cdt - ∇.Lv) + (dLv/cdt + ∇Ls + ∇xLv)

or work = XL= (dhs/dt - ∇.Lv) + (dhv/dt + ∇Ls + ∇xLv)

Einstein's Photoelectric Equation is the scalar part of (6). Einstein called work the kinetic energy and the divergence of the vector Lv, he called the "work function, phi". Planck had previously called dhs/dt, hsf the quantum of energy. It was Einstein's Photoelectric explanation incorporating Planck's Quantum Hypothesis, that gave life to Quantum Theory as many physicists saw no basis for Planck's hypothesis other than it fit the facts, which was the reason Planck invented it, to fit the facts! Einstein's Photoelectric Theory at the same time gave support to Newton's Particle Theory of Light , solidifying the Wave-Particle Duality in physics. From these small beginnings, Quantum Theory went on to dominate physics.

B. Relativity Theory

As opposed to Work or Energy, Einstein's Theory of General Relativity (Einstein [2]) focuses on force, especially Gravitational force. Einstein genius was too recognize and accept the obvious reality of Nature. He accepted the constancy of the speed of light and the symmetry of motion to craft Special Relativity. He also accepted the identity of inertial force and gravitational force to craft General Relativity. Like Planck, his explanation fit the facts. Space was curved by Gravity and "matter" created gravity. His famous "Cosmological Constant" was introduced to account for the quasi- static distribution of matter, as required by the fact of small velocities of the stars. In the Theory here proposed Force, including Gravity, is shown to be the second derivative, or curvature of Life L. These equations show the universe force to be in general hyperbolic.

(7) $F = X^2 L = ((T^2 - \nabla^2)Ls - 2T\nabla.Lv) + ((T^2 - \nabla^2)Lv + 2T(\nabla Ls + \nabla xLv))$

Force is also a quaternion consisting of a scalar longitudinal wave and a vector transverse wave. The scalar manifests Gravitational effects and the vector Electromagnetic effects. This explains the Wave-Particle Duality of physics as 2 waves, a scalar wave and a vector wave. The wave nature is the result of the Curvature Operator, X^2. Einstein's Cosmological Constant can be seen in the scalar longitudinal wave, to be equal to $2T\nabla.Lv$, or the time derivative of the divergence of Lv.

(8) $X^2L = ((T^2 - \nabla^2) + 2T\nabla)(Ls + Lv)$

Curvature of Space means mathematically that Force F is non-zero. Positive force is like looking into a Pit, Negative force is like looking up at a Peak, Zero force is like looking at a flat Plane. No Force means No Curvature means the Universe is Flat like a Plane and represents Equilibrium. All three possibilities would appear to exist. Sections of the Universe could be curved positive, negative or flat. Another question is Does the Universe have a Change boundary, a limit? Does 0 = XL. If it does then Force becomes:

(8 a) $F = - (T^2 + \nabla^2)L$

This limit however, could be finite or infinite depending on the value of L at 0 = XL. If L is finite Force is finite and the universe is finite. If L is infinite, the Universe is infinite. In either case Force at the limit is seen to be elliptical, not hyperbolic. Limits are local and could vary at different times and places. To answer the question, Does the Universe have a limit, I turn to Quantum Theory. If 0 = XL, then Life is Conservative, has a boundary, a limit. Planck's Hypothesis hsf = Quantum Energy E is equivalent to 0 = XL at least for the scalar part of XL in equation (6). Planck never conceived of a Action vector, hv or the resulting vector part of (6). Einstein's Photoelectric Equation did not conclude that 0 = XL. Einstein explained that Electron Kinetic Energy = hsf - ∇ .Lv. Who is correct here? The vector part of the equation 0 = dhv/dt + ∇ Ls + ∇ xLv, is not addressed by either Planck or Einstein or others. I hold that the vector equation is the Quantum equivalent of Lenz's and Newton's Law of Action and Reaction. This would mean that the vector radiation dhv/dt plus the gradient ∇ Ls and the curl ∇ xLv are DEPENDENT on each other and sum to zero. Quantum Theory to date does not address this vector equation. However, Electromagnetism does and it is the link to Electromagnetism and light that leads me to conclude that Life is Conserved, $0 = XL$, and Planck's scalar equation reflects the quaternion equation $0 = XL$. This we take up with the investigation of Maxwell's Equations and Electromagnetism. Planck and Einstein's Equations are both correct, the condition $0 = XL$ defines the Change Boundary of the function L. Internal to that boundary XL is non-zero. Einstein's Photoelectric equation is a non-boundary condition.

5. Maxwell's Equations Revised

Maxwell's Equations reflect the state of experimental physics in the 1800's. Maxwell aimed at putting a sound mathematical foundation under the work of primarily Faraday. Maxwell's Great Achievement was to predict that Light was electromagnetism by noting the relationship between electrostatic and electromagnetic units! This relationship was a constant velocity with value c, the speed of light. It was in paying attention to elementary units that Maxwell made one of the Greatest discoveries in science (Maxwell [3]). In his mathematics , Maxwell used Hamilton's concepts of vectors and vector operator ∇. Maxwell had problems with Hamilton's Rules for vectors $I^2 = J^2 = K^2 = -1$. These Rules often gave a minimum where physicists were expecting a maximum. For example Physicists wanted a falling ball to have positive energy and Hamilton's Rules said the energy is negative or exergy not energy. Consistent mathematics was sacrificed to confused conventions of the 19th century. Convention conquered Consistency and Gibbs and Heaviside invented Vector Analysis to replace Hamilton's Quaternion. Vector Analysis became the foundation of physics and mathematics rather than Quaternions after 1900 even though Quaternions form a Division Algebra and Vector Analysis does not. Vector Analysis also does not satisfy Closure or Associativity.

Maxwell's Equations deals with the Electric Intensity Field E, the Magnetic Intensity Field H, The Electric Density Field D and the Magnetic Density Field B. Maxwell related charge density ρ to the Divergence of the Electric Density vector $\nabla.Dv$ but concluded that there can be no magnetic equivalent to ρ because $0 = \nabla.Bv$. This conclusion is dubbed the "No Magnetic Monopole" and was based on the lack of observation of a Magnetic Monopole, and on a supposed mathematical and theoretical difference between the Ev-field and the Bv-field. Ev was assumed to be the Gradient of a scalar ∇Es and Bv= $\nabla x Av$, the Curl of a Vector field. This by definition would make $0 = \nabla.Bv$, from the identity $0 = \nabla.\nabla x Av$, which holds for any vector Av. Ampere's work defining Bv as the result of currents brought the Theory and experiments together into a conclusion of No Magnetic Monopoles. The Theory here presented is that Electromagnetism reflects the Conservation of the Quaternion Electric Field E. The Change Operator is the same 4 dimensional Quaternion Operator $X = T + \nabla$:

Maxwell's Equations Revised are the result of the Stationary E field, $0 = XE$:

(9) $0 = XE = (dEs/cdt - \nabla.Ev) + (dEv/cdt + \nabla Es + \nabla xEv)$

or $0=XE = (dBs/dt - \nabla.Ev) + (dBv/dt + \nabla Es + \nabla xEv)$

This gives two of Maxwell's Four Equations, a scalar and vector equation:

(10) $0 = dBs/dt - \nabla.Ev$ and $0 = (dBv/dt + \nabla Es + \nabla xEv)$

Multiplying thru by y, the admittance y, gives the other two equations, $H = yE$:

(11) $0 = XH = ((dHs/cdt - \nabla.Hv) + (dHv/cdt + \nabla Hs + \nabla xHv)$

or $0=XH = (dDs/dt - \nabla.Hv) + (dDv/dt + Jc + \nabla xHv)$

Note that Jc is the conduction current equal to ∇Hs, the gradient of the scalar Hs. The scalars seem to account for the "matter/charge" effects. However, this Quaternion Theory is a Pure Field Theory, with no explicit "matter/charge". This can be best seen in by dividing equation (11) by the scalar c to give the D field:

(12) $0 = XD = (dDs/cdt - \nabla.Dv) + (dDv/cdt + \nabla Ds + \nabla xDv)$

or $0 = (\rho s - \nabla.Dv) + (\rho v + \nabla Ds + \nabla xDv)$

This equation indicates ρ the charge density, is the time derivative of D, divided by c, or the scalar derivative of D.

The impossible "Magnetic Monopole, $0 = \nabla.Bv$" is seen to exist wherever there is charge density and impedance and is equal to $z\rho$ or the impedance z times the charge density ρ, from $0 = dBs/cdt - \nabla.Bv$ gives $dBs/cdt = dzDs/cdt = z \rho$.

(12 a) $0 = zXD = XB = (dBs/cdt - \nabla.Bv) + (dBv/cdt + \nabla Bs + \nabla xBv)$

Maxwell's Equations can be seen to be but one quaternion equation with the E,B,H and D fields all being related by c and zo = voltsec/charge (flux/charge) . Maxwell's Ether I believe exists and is manifested by "free space" impedance zo. The square of the Ether's magnetic flux is given by $m^2 = zoh$, zo times Planck's constant h. The Ether's charge squared is given by $e^2 = h/zo$.

6. Continuity of Life: 0 = XL

Planck's Quantum hypothesis hsf equals Quantum of Energy (∇.Lv), indirectly coupled radiation to "matter". Planck as most physicists defined energy to be a scalar and gave a different name to the vector complement of energy, torque. Thus they avoided coming to grips with quaternions. Einstein's Photoelectric Equations directly coupled radiation to "matter" with his work function phi. However, Einstein did not investigate the vector work. My hypothesis is that the Quantum Hypothesis is "Continuity of Life" :

(13) $0 = XL = (dLs/cdt - \nabla.Lv) + (dLv/cdt + \nabla Ls + \nabla xLv)$

or $0 = XL = (dhs/dt - \nabla.Lv) + (dhv/dt + \nabla Ls + \nabla xLv)$

This hypothesis is essentially Planck's Hypothesis, hsf = ∇.Lv:

(14) $0 = XL$ or $0 = (dhs/dt - \nabla.Lv)$ and $0 = (dhv/dt + \nabla Ls + \nabla xLv)$

This equation (14) states essentially the Continuity Condition. Scalar Continuity is $0 = dhs/dt - \nabla.Lv$ and Vector Continuity is $0 = (dhv/dt + \nabla Ls + \nabla xLv)$. Another interpretation is that of Dependency. The scalar and vector terms are interdependent. A final interpretation is that of Boundary Conditions, the place where Change no longer happens, or the conditions, which preclude change. The vector Continuity Equation confines the three vectors to a plane. The three vectors could form a triangle, be termino co-linear, co-linear or all null.

I take support for this Stationary Hypothesis from many areas. First is Electromagnetism. The behavior of Electromagnetic fields seems to support Maxwell's Equations in "free space". " Free Space" is the condition of:

(15) $0 = XEs = dEs/cdt + \nabla Es$

This is the condition of the Ether, no "matter/charge" change. It is not necessary that $0 = Es$, only that there be no Change in Es. Maxwell's Equations had trouble with "matter/charge" and necessitated the invention of the electron as an extra field concept. Using quaternion E fields naturally accounts for "matter/charge" effects. This is seen in $\rho = \nabla.Dv$ and Jc in the non-free space equations. However, these terms are appended and not derived as here from a Pure Field Theory, E = Es + Ev.

Another confirmation is the vector Continuity equation in electromagnetism:

(16) $0 = dBv/dt + \nabla Es + \nabla xEv.$

This explains Lenz's and Newton's Law or "every action has an equal and opposite reaction". The gradient of Es, DEs, gives the so-called "back-emf", talked about but never directly defined. The combination of Planck's Quantum Theory and the behavior of Electromagnetic Fields, lead me to the belief that Life is Stationary or Immortal in the Universe. The electromagnetic and gravitational fields are manifestations of Life. Force is the Curvature of Space:

(17) Force F = eE = X^2L

Because Life is Stationary, $0 = XL$, force at the boundary is Elliptical:

(18) Force F = - $(T^2 + \nabla^2)L$

Depending on the sign of F, the universe is positively curved (minimum), negatively curve (maximum) or flat at 0, and inflection point. The hyperbolic waves exist only internally to the Boundary of the Universe at non-stationary points.
The Cosmological Constant exists only internally with hyperbolic waves. Internally, the universe can exist in hyperbolic equilibrium $0 = X^2L$, at the boundary equilibrium is elliptical $0 = (T^2 + \nabla^2)L$. At equilibrium or Curvature Boundary, force is zero, no curvature, a flat condition of the universe. Force disappears at $0 = X^2L$.

(19) $0 = X^2L = ((T^2 - \nabla^2)Ls - 2T\nabla.Lv) + ((T^2 - \nabla^2)Lv + 2T(\nabla Ls + \nabla xLv))$

The Gravitational Force Wave under these conditions is zero and:

(20) $((T^2 - \nabla^2) Ls = 2T\nabla.Lv$.

The Electromagnetic Force Wave is also zero:

(21) $0 = ((T^2 - \nabla^2)Lv + 2T(\nabla Ls + \nabla xLv))$.

7. Conclusions

Quaternions numbers and operators are the natural analysis tools of physics. Using these tools the unification of apparently irreconcilable Theories may be possible. The Ether may be restored through scalars and analysis simplified by a powerful but simple mathematical tool. The "free space" impedance zo may be a measure of the Ether. zo has the units of 377 Ohms and is the ratio of the "free space" flux m to the "free space" charge e. The product of hs and zo gives the flux squared. The flux can then be calculated and is about .5 femto voltseconds! The ratio of hs/zo is the charge e squared, the value of e is 1.326 atto coulombs or about 8.25 electrons. Maxwell's Ether pervades the Universe and is detectable in the "free space" electromagnetic impedance, zo!

Life is All and All is Life. Leben ist Alles, und Alles ist Leben.

I. Mathematical Appendix:

A brief introduction to Quaternion is presented here. Quaternions are not widely used in modern mathematics and Physics, though their use is spreading in the communications and graphics industry.

Hamilton's Rules for Quaternions distinguish Quaternions from Vector Analysis created by J. Willard Gibbs. Hamilton' Rules enable Quaternions to satisfy the properties of a Group for Addition and Multiplication.

These Group properties are :

1. Closure {A, B and AB are members of the group}
2. Identity { For Identity I, AI = A}
3. Associativity { A(BC) = (AB)C }
4. Inverse { For Inverse A^{-1}, A A^{-1} = $A^{-1}A$= I}
5. Commutivity { AB = BA holds for scalars only }

Types of Quaternions:

L = Ls + Lv where Ls= Lo is the scalar part and Lv = IL1 + JL2 + KL3 is the vector part.
L0,L1,L2 and L3 coefficients can be complex numbers, a + ib, where i is the scalar square root of -1.

The Conjugate of L is L' = Ls- Lv

The Norm of L is NL = LL' = L'L = $L0^2$ + $L1^2$ + $L2^2$ + $L3^2$

The Inverse of L is L^{-1} = L'/NL

The Unit Quaternions have much to do with our "Elementary Particles" . If this is true, then there are 320 "Elementary Particles".

There are 80 real Unit Quaternions with Norm = + 1:

8 Singlets: +/- (1,I,J,K)/1(square root of 1)
24 Doublets: e.g. (+/- 1 +/- I)/square root of 2
32 Triplets: e.g. (+/- 1 +/- I +/- J)/square root of 3
16 Quadruplets: e.g. (+/- 1 +/- I +/- J +/- K)/2 (square root of 4)

There are 80 imaginary unit Quaternions with Norm = - 1.

There are 80 complex unit Quaternions with Norm = +i : e.g. ((1 +i)/square root of 2) times {+/- (1,I,J,K)}

There are 80 complex conjugate Quaternion with Norm = -i : i.e. ((1 -i)/square root of 2) ,

Vector Differential Identities:

0=$\nabla.\nabla$xLv
0=∇x∇Ls
$\nabla(\nabla.Lv)$ = ∇^2Lv + ∇x∇xLv

10

J. Footnotes and references

1. ELEMENTS OF QUATERNION BY WILLIAM ROWAN HAMILTON, CHELSEA PUBS NY, NY 1969
2. THE PRINCIPLE OF RELATIVITY BY A. EINSTEIN, H. LORENTZ, H. MINKOWSKI AND H. WEYL, DOVER PUBS 1923
3. A TREATISE ON ELECTRICITY & MAGNETISM, BY JAMES CLERK MAXWELL, DOVER PUBS NY, NY 1954
4. Seven Ideas that Shook the Universe BY Nathan Spielberg and Bryon Anderson, JOHN WILEY PUBS, CANADA, 1987

Homeostasis
© 2001 by Wardell Lindsay

ho-me-o-sta-sis (ho|me o|stasis; *also* , -stasis) *n.* [[ModL: see HOMEO- & STASIS]] **1** *Physiol.* the tendency to maintain, or the maintenance of, normal, internal stability in an organism by coordinated responses of the organ systems that automatically compensate for environmental changes **2** any analogous maintenance of stability or equilibrium, as within a social group --**home-o|static** *adj.* [1]

Claude Bernard, a French Physician, introduced the concept of Homeostasis in his practice of medicine. This same principle or rule, applies to the operation of the Universe. This is an example of the power of studying science. Newton's Falling Apple, is scientifically no different from the falling Moon that orbits our Earth. To understand one, is to understand the other.

Change is the central concept of Life and science. As we individually live, we individually change. As we as a species live, we also change and call this change evolution. Darwin's Theory of Evolution ('On the Origin of Species by Means of Natural Selection', which appeared in 1859.), stated that we change by means of "Natural Selection", which includes external environmental factors like geography and climate and internal factors like genes.

Darwin's Theory does not deal with "Origin" of species, despite the title of his book. The more correct title would have been the "Change of Species by Means of Natural Selection". It could be argued that Change is a form of Origin, but this would make evolution disappear as a Theory, because each evolution of a species would be an "original" species. It is the connection of Origin with Original that makes Darwin's Title or his Theory wrong. The Theory of evolution by Means of Natural Selection is substantially correct by scientific standards, (supported by evidence consistent with the theory). The Origin in the Title is scientifically unproven., not the same as incorrect, but not substantiated by evidence, the "missing link" is still missing.

Creationism, an alternative to the Origin, is consistent with the Theory of Evolution, but differs on the Origin Question. Most scientists believe the mechanism of Evolution could be responsible for the Origin and as such make Darwin's Theory complete. Many people, believe the Creator could also be responsible for the Origin. In the history of mankind, there have been three answers to the role of the Creator. In very ancient times, mathematics the Greek word for Learning, was a subset of God's Creation. God created everything including mathematics or learning. God is bigger than Learning or Mathematics. Among early peoples, the ancient Egyptians, a very learned society, developed the idea of Maat to mean that God and Learning or Mathematics are the same. Finally in Western or Modern times, the idea has become current that Mathematics or Learning (Science) is a truth that is larger than God in the sense that God cannot suspend or change the rules of mathematics. [PI IN THE SKY by John D. Barrow, page 256, 1992, Little, Brown and Company].

My own view is that the Egyptians are correct. Maat the Creator is the same as mathematics or learning. The Egyptian concept of Maat, includes Order, Justice and

---

[1] From Compton's Interactive Encyclopedia © 1998 The Learning Company, Inc.

Balance. The Egyptian Maat, is the origin of our concept of Justice, the lady with the Scales, and the concept of Judgment after Death.

Life, Evolution and the Universe is based on the concept of Change, to become different. Different is with respect to something . In science, it is different with respect to time and space. This difference is related to the environment in which we live, a time and a place. Mathematically, we live in a 4 dimensional universe. Our universe has one dimension of time (ct, where c is the speed of light) and three dimensions of space (Ix, Jy and Kz). The three dimensions of space are vectors dimensions denoted by the I, J and K vectors. The time dimension is a scalar dimension. The scalar (ct) time dimension is omni-directional, all directional.

The ancient Egyptians had a god for change called Xepra. Xepra was symbolized by the Beetle who "transformed " a little ball of dung into a new beetle. To the Egyptians change was a transformation activity. Change is expressed mathematically by the ratio of something to the four dimensions (3 space and 1 time). I have created a mathematical operator I call X, to express four-dimensional change. Historically, William Rowan Hamilton, the Irish mathematician and Physicist created the concept of vectors when in 1843, he invented quaternions , a scalar and three vectors. Quaternions were created in Hamilton's quest to mathematically rotate a line or vector in three space. He found that it required a scalar and three vectors to accomplish this. Thus the creation of quaternions. Hamilton also created the vector change operator Del (∇ = Id/dx + Jd/dy + Kd/dz). These developments by Hamilton revolutionized mathematics and physics. Quaternions introduced the concept of non-commutative algebra (AxB does not equal BxA). Vectors behaved differently from the prior numbers which were scalars, (ab= ba).

Newton and Leibniz had previously invented a scalar change operator as the change operator in calculus. This was the time derivative operator d/dt. Calculus was the mathematics of scalar change and Vector Calculus was the mathematics of vector change. Hamilton, while he created a four dimensional algebra in quaternions and a three-dimensional vector change operator called Del ∇, Hamilton did not create a four-dimensional change operator.

Around 1965, while studying electromagnetism, I created a four-dimensional change operator ,

$$X = d/cdt + \nabla = d/cdt + Id/dx + Jd/dy + Kd/dz = T + \nabla$$

X is a quaternion differential operator consisting of a scalar d/cdt and a three vectors ∇. Mathematically, X provides a calculus compatible with the four dimensional space-time of the universe. I believe and wish to state that the universe is quaternionic, that is , the universe consists of scalars and vectors . The fundamental variables of the universe, work , energy, force, momentum, etc are quaternions . I believe the universe is fundamentally a field or "spiritual" entity, rather than a substantive/matter entity. For example Einstein's $E = mc^2$, says there is a relation between m-matter and E -energy. Is matter a form of energy or is energy a form of matter? I choose to say that "matter" is a

3

form of energy, specifically, matter is Energy/c^2 ! I will not dwell on this topic here, but you can see the potential for confusion.

Once we have a framework for the universe, space-time and change X, we can proceed to define the fundamental variables or constructs. Physics and the universe can be described by any number of fundamental concepts. Before electricity, the universe was described in terms of forces. These forces were mostly mechanical, then gravitational, then thermal, then electrical and now nuclear. The concept of force gave birth to energy which was seen to be more fundamental and related to force by force = energy/displacement. The displacement is a vector concept because force was originally a vector and energy a scalar. A vector times a scalar gives a vector. However, a vector times a vector can give a scalar or a vector depending on the angle between the two vectors. If the angle is 0 or 180 degrees, then a scalar results; if the angle is an odd multiple of 90 degrees a vector results; otherwise a vector and a scalar results. Quaternions, greatly simplify this situation, in that in all cases a quaternions results, a concept called mathematical Closure. Vectors are not mathematically Closed. Quaternions are!

The fundamental variable in the universe I call Life L=Ls + Lv. Life is a quaternion consisting of a scalar part Ls and a vector part Lv=ILx + JLy + KLz. Life L is four-dimensional 1 scalar Ls and three vectors Lv. Life has units of energy-feet. [The units for energy can be joules or Newton-meters or pound-feet or pound-mile or electron volts or calories.] The advantage of the variable L, is that the name Life brings to light the fact that the operation of the universe is life-like, if not identical to life or living. I believe the universe is alive and is immortal.

With Life as the fundamental variable of the universe, the first law of the universe is change. The change of Life is energy or work W:

$$W = XL = (dLs/cdt - \nabla.Lv) + (dLv/cdt + \nabla Ls + \nabla x Lv)$$

For Life to be immortal, there can be no change or W must be zero! When change is zero this means the components of change are dependent on each other, such that if one component increase another decreases in just such a fashion to result in no overall change - remember homeostasis. For W to be zero, both the scalar part (dLs/cdt - ∇.Lv) must equal 0 and vector part (dLv/cdt + ∇Ls + ∇xLv) must be zero.

$$0= (dLs/cdt - \nabla.Lv) \text{ and } 0 = (dLv/cdt + \nabla Ls + \nabla x Lv)$$

This is the mathematical statement of Homeostasis !
The Homeostasis equation derives mathematically Planck's Quantum Theory. Planck's Quantum Theory is the result of an explanation for the failure of Maxwell's Electromagnetic Theory to explain radiation. To account for the actual distribution of radiant energy, Planck assumed that energy was "quantized" by frequency E=hf, where E is the quantum energy , "h" is Planck's action constant and f is the frequency. Action has units of energy-seconds. Action (E-s) times frequency (1/s) gives energy. Planck's

Equation is derived when it is recognized that Life has units of energy-feet and that L divided by the speed of light c gives action h. Thus the homeostasis equation :

$0 = (dLs/cdt - \nabla.Lv)$ and $0 = (dLv/cdt + \nabla Ls + \nabla x Lv)$

transforms to :

$0 = (dhs/dt - \nabla.Lv)$ and $0 = (dhv/dt + \nabla Ls + \nabla x Lv)$

Where $dhs/dt = hsf$ and $E = \nabla.Lv$

The vector Homeostasis Equation $0 = (dhv/dt + \nabla Ls + \nabla x Lv)$ is not currently a part of Quantum Theory but explains the interaction of vector radiation (dhv/dt) , the Gradient of "matter" (∇Ls) and the interaction of the curl of the vector field ($\nabla x Lv$). These same fields are seen in the equations of electromagnetism and described by Lenz's Law. The general description of the vector Homeostasis equation is given by Newton's saying "For every Action there is an equal and opposite re-action". This is the meaning of the vector Homeostasis equation.

Maxwell's Equations are meant to describe the interaction of electromagnetic fields. They are generally correct for fields in "free space or no matter". The correct equations for electromagnetism can be derived form the Homeostasis Equation for the electric field E,

$0 = (dEs/cdt - \nabla.Ev)$ and $0 = (dEv/cdt + \nabla Es + \nabla x Ev)$

recognizing that $E = cB = zH = zcD$. Gives two of Maxwell's Four Equations:

$0 = (dBs/dt - \nabla.Ev)$ and $0 = (dBv/dt + \nabla Es + \nabla x Ev)$

Maxwell's other two equations come from substituting $E = zH$ or

$0 = (dHs/cdt - \nabla.Hv)$ and $0 = (dHv/cdt + \nabla Hs + \nabla x Hv)$

recognizing that $E = cB = zH = zcD$. Gives the other two of Maxwell's Four Equations:

$0 = (dDs/dt - \nabla.Hv)$ and $0 = (dDv/dt + \nabla Hs + \nabla x Hv)$

Maxwell's Equations can be corrected with the equations presented here.
The other major variable in the universe is Force. Force is given by the second derivative of change , which creates curvature and wave equations:

$X^2 = ((d^2/c^2 dt^2 - \nabla^2) + 2d/cdt \nabla) = ((T^2 - \nabla^2) + 2 T \nabla)$

Force $= F = X^2 L = ((T^2 - \nabla^2) Ls - 2T \nabla.Lv) + ((T^2 - \nabla^2) Lv + 2T(\nabla Ls + \nabla x Lv))$

5

Force is also a quaternion with a scalar part and a vector part. The scalar wave is longitudinal and the vector wave is transverse. These vector wave account for the so-called Wave-Particle Duality of Light". The longitudinal wave accounts for the "Gravity" wave.

Einstein's now popular "Cosmological Constant" can be seen in the scalar wave to be the time derivative of the divergence, 2T ∇.Lv. Einstein was correct in disowning the constant, it is not a constant.

Here we see again the convenience and fundamental importance of the Life variable as it derives the two most important factors in physics energy and force, from the change operator alone.

When force is zero, the universe is flat or without curvature and the waves are hyperbolic.

Under Homeostasis or at the Boundary, the Force Equations is elliptical:

Force = F= - ($T^2 + \nabla^2$) L

Einstein's Theory of Relativity is reflected in the Force Equations. The gravitational Force is the scalar part and the electromagnetic Force is the vector part. "Matter" is a manifestation of scalars in a quaternion Theory of Physics, allowing for a "Pure Field" Theory of Physics as advanced by Faraday, Maxwell's Mentor.

Maxwell's Ether. Maxwell believed that electricity carried a form of energy and that this energy was transmitted from the source to the destination thru the ether. He believed that between the source and the destination, the energy traveled in the ether. Does this ether exist? Examining the constants of Quantum and Electromagnetism, I believe the ether exists! Planck's Quantum action Constant h = MC, is the product of two ether constants a magnetic flux M in voltseconds and charge C in coulombs. Are there other constants containing these two constants? The answer is yes, the characteristic of the ether itself, the impedance of "free space", zo! The "free space " impedance is zo=M/C. We now have two unknowns and two equations and simple algebra allows us to solve for the constants of the ether. Solving these two equations where h= 663 atto atto joule seconds and zo= 377 ohms gives the constants of the ether:

M = 500 atto voltseconds and C = 1.326 atto coulombs. = 8.25 electrons

ElectroMagnetism
Copyright 2002 by Wardell Lindsay

1. The fundamental forces of Physics and nature are electromagnetic. From Radio to Radar to Nuclear Reactors, electromagnetism rules. Laws governing Electromagnetism were first discovered in the 1820's. Since then, Electromagnetism has evolved but the Laws are incomplete. Maxwell in the 1860's made significant progress in identifying light as electricity and developing an equation describing electric waves. His equations are the basis for our current understanding, but they provide only a partial description of the Laws of Electromagnetism. The complete description will be developed here.

2. To understand Electromagnetism, you must understand change. In nature there are two kinds of change. Change with respect to time and change with respect to space. For most of the history of science these two kinds of change did not seem to have any necessary connection. Calculus is the mathematics focusing on change and calculus treated change with respect to time separately from change with respect to space. To understand Electromagnetism, there needs to be a new calculus combining time and space change.

The simple addition of a time derivative with a space derivative answers this need. I developed such a change operator called Xepera X, consisting of a time derivative $T = d/cdt$ and a vector space derivative. The speed of light c is a scalar and as such is omni-directional. It is a constant in the universe and converts the time dimension, t, into space units in the X, the change operator.

[1] Xepera X= T + ∇ = d/cdt + Id/dx + Jd/dy + Kd/dz = d/cdt + ∇

3. The time derivative is a scalar meaning it is omni-directional and the space derivative is a vector, meaning it is directional. The space directions are indicated by the vectors, I, J and K. These vectors denote the x, y and z directions. Vectors multiplication is ordered, meaning IJ = - JI. = K Order of multiplication counts. The rules governing vectors were given by their developer William Rowan Hamilton and can be summed up in what I call Hamilton's Vector Rules:

[2] II² = J² = K² = IJK = - 1 and IJ=K, JK = I and KI = J

4. Hamilton developed the mathematics of scalar and vectors and called them Quaternions. Quaternions are a four-dimensional mathematics consisting of one scalar and three vectors. For example the quaternion E-field has a scalar Es and three vectors:

[3] E = Es + IEx + JEy + KEz = Es + Ev

Science today largely ignores the scalar fields and focuses on the vector fields. Scalars are in my theory very important and represent "matter and charge" effects in electromagnetism. Quaternions are mathematically very elegant, powerful and complete. Mathematically speaking, quaternions contain scalars, which include the complex numbers, and vectors. In addition quaternions are the only Division Algebra. Real number, complex number and quaternion division are all subsets of quaternions.

5. There are four fields of interest in electromagnetism and they are all related by c the speed of light and z the "free space" impedance.

[4] E = cB = zH = zcD

E is the Electric Intensity field with units Volts/foot
B is the Magnetic Density field with units Weber/foot or Volt-seconds/foot^2
H is the Magnetic Intensity filed with units Amperes/foot
D is the Electric Density with units of Coulomb/foot^2

6. The free space impedance z reflects the resistance to electromagnetism provided by the ether or "free space". "Free space" has been thought to be free of "charge or matter", a vacuum, and empty, yet there is electromagnetic resistance there. This electrical contradiction has led me to investigate and conclude that: free space" is not a vacuum, but contains electric and magnetic charge! The charge is scalar and is a constant. The impedance z is equal to the ratio of the magnetic and electric charge:

[5] Z = m/e where m is the magnetic charge and e is the electric charge

The value of the z is known to be 120 Π Ohms or 377 Ohms. Finding the value of m and e requires another constant in nature. The close relation between electromagnetism and quantum theory led me to investigate Planck's Action Constant h. This proved fruitful and assuming that h is also related to the electric and magnetic charges by h= me, the value of m and e can be found. The values are in atto units or 10 to the minus 18 parts

[6] M = 500 atto webers and e = 1.326 atto coulombs

This finding means the space of the universe is filled with electric and magnetic charges reflected in the resistance to electromagnetic waves and provides the means for propagating the electromagnetic fields.

7. With this introduction the Laws of Electromagnetism can be found. Change Happens. If change did not happen, Life and the Universe would be dull. Change is the basis for evolution. Change is the product of XE and is illustrated in the table below.

XE =	Es	IEx	JEy	KEz
d/cdt	dEs/cdt	IdEx/cdt	JdEy/cdt	KdEz/cdt
Id/dx	IdEs/dx	-dEx/dx	KdEy/dx	-JdEz/dx
Jd/dy	JdEs/dy	-KdEx/dy	-dEy/dy	IdEz/dy
Kd/dz	KdEs/dz	JdEx/dz	-IdEy/dz	-dEz/dz

The law of electromagnetism, however, is the law of Constancy or no Change:

[7] $0 = XE = (dEs/cdt - \nabla.Ev) + (dEv/cdt + \nabla xEv + \nabla Es)$

This equation expresses the Law of Conservation, Constancy and Dependency. This says there is no change in the Quaternion Electric Field, This means the scalar part is zero and the vector part is zero:

[8] $0 = (dEs/cdt - \nabla.Ev)$ and $0 = (dEv/cdt + \nabla xEv + \nabla Es)$

The scalar equation $0 = (dEs/cdt - \nabla.Ev)$ is called the Continuity Equation.
The vector equation $0 = (dEv/cdt + \nabla xEv + \nabla Es)$ is called the Stability Equation

The traditional equations are easily derived by recognizing that E =cB or B=E/c. This gives

[9] $0 = (dBs/dt - \nabla.Ev)$ and $0 = (dBv/dt + \nabla xEv + \nabla Es)$

The other notable equation is given by

[10] $0 = XH = (dHs/cdt - \nabla.Hv)$ and $0 = (dHv/cdt + \nabla xHv + \nabla Hs)$

Substituting D= H/c gives the traditional terms

[11] $0 = (dDs/dt - \nabla.Hv)$ and $0 = (dDv/dt + \nabla xHv + \nabla Hs)$

In the experimental work of Oersted, Faraday and Ampere and the mathematical work of Maxwell and Gauss, equations similar to these equations were developed and called **MAXWELL"S EQUATIONS**, the most famous equations in physics.
They are presented here for comparison:

[12] $0 = (\rho/\varepsilon - \nabla.Ev)$ and $0 = (dBv/dt + \nabla xEv)$

[13] $0 = (0 - \nabla.Bv)$ and $0 = (dDv/dt - \nabla xHv + J)$

8. The use of quaternion calculus led to the simple development of the mathematical Laws of Electromagnetism. It also showed the role and importance of the scalar fields and their relation to "matter and charge". It is also clear that the fundamental constants are c, z and h. Physical "matter" represents the scalar fields for example the conductors and Es represents antennas in electric circuits. Ev represents the vector fields. The Law of change shows how the scalar and vector fields relate to each other.

The vector laws of associativity are:

[14] $0 = \nabla.\nabla XEv$ $0 = \nabla x\nabla Es$ and $\nabla(\nabla.Ev) = \nabla x\nabla xEv + \nabla^2 Ev$

The law of Change X operates on itself to give Curvature:

[15] Curvature $X^2 = (T^2 - \nabla^2) + 2T\nabla = ((d/cdt)^2 - \nabla^2) + 2T\nabla$

Curvature is the second derivative of Change in Spacetime. Curvature provides the Wave Equations of Electromagnetism. Maxwell had predicted that light was a form of Electromagnetism and developed a wave equation:

[16] $0 = (T^2 - \nabla^2)Ev$

The quaternion wave equation is given by the Curvature:

[17] $X^2 E = ((T^2 - \nabla^2)Es - 2\nabla.dEv/cdt) + ((T^2 - \nabla^2)Ev + 2(\nabla xdEv/cdt + \nabla dEs/cdt))$

The quaternion wave equations consist of a scalar wave and a vector wave. The scalar is longitudinal (divergence of vibrations) and "gravity like " and the vector wave are transverse (Curl of vibrations) and electromagnetic like. This wave denotes the transverse and longitudinal nature of the electromagnetic waves mathematically and explicitly. The scalar coefficients can be complex and reflect the polarization of the waves, imaginary coefficients for vertical polarization and real coefficients for horizontal polarization.

Curvature at the Boundary Condition, or Constancy Condition is given by:

[18] $X^2E = - (T^2 + \nabla^2)E$

Space is said to be flat when Curvature is zero

[19] $0 = X^2E = ((T^2 - \nabla^2)Es - 2\nabla.dEv/cdt) + ((T^2 - \nabla^2)Ev + 2(\nabla xdEv/cdt + \nabla dEs/cdt))$

9. The fundamental constants are intrinsic to Electromagnetism:

[20] Webers m $= \int \nabla.Bv \ dV = \int Bv.dS = \int Bv.dA$

[21] Coulomb e $= \int \nabla.Dv \ dV = \int Dv.dS = \int Dv.dA$

[22] Voltage V $= \int \nabla xEv.dA = \int Ev.dc$

[23] Ampere A $= \int \nabla xHv.dA = \int Hv.dc$

[24] $z = m/e$

[25] $h = me$

"The Universe is Alive!"
©2002 by Wardell Lindsay

Introduction

This work is in honor of the Ancient Egyptians who were brilliant and believed that we are all part of the Living Universe and Life continues after death. My work is a mathematical/physical Theory uniting Einstein's Relativity Theory and Planck's Quantum Theory in a Theory of the Living Universe.

Life is a concept related to "Action" in physics and is here the content of the Universe. The content Life, is contained in the Universe by a container we call Spacetime. Spacetime is four-dimensional container, with three space vector dimensions (Ix, Jy, Kz) where I, J, and K denote space vectors and the fourth scalar dimension (ct), where c is the speed of light and t is the time dimension! Spacetime is measured in units of length, not time. The symbols "t, x, y and z" represent distance from an assumed origin in any given calculation. It is not known whether there exists a "real origin, where "t, x, y and z" all have value zero at this real origin. For example, the Sun is assumed as the "origin or center " of our solar system and the Earth is assumed as the "origin" in some satellite work. A point in Spacetime can be described by a four dimensional Point P= ct + Ix + JY + Kz. The point P can be simply described as P= Ps + Pv, where Ps =ct is the scalar part and Pv =Ix + Jy + Kz is the vector part. William Rowan Hamilton in the 1843 called points like P, the sum of a scalar and three vectors, quaternions.

Hamilton thought very highly of quaternions, believing they would make major contributions to Physics. Having invented the concept of quaternion consisting of scalars and vectors, Hamilton's vector product (cross x) was accepted and hailed as the first Non-Commutative Algebra, meaning IJ = - JI. = K This has the effect of making the order of quaternion multiplication important, as AB does not equal BA, unless the vector parts are parallel. Hamilton had two kinds of multiplication for vectors. Scalar multiplication designated here by a dot "." and vector multiplication for vectors designated here by a cross "x". Hamilton's scalar product (dot.) rule for squaring vectors ($I^2 = J^2 = K^2 = -1$) was rejected by physicists and replaced by Josiah Willard Gibbs's Vector Calculus with the Rule for squaring vectors ($I^2 = J^2 = K^2 = +1$). This seemingly simple change of sign makes Hamilton's Rules preserve mathematical associativity:

$(I^2) J = I (IJ) = -J$ and Gibbs's Rules doesn't: $I^2 J = +J$ and $I (IJ) = -J$.

In the 1800's four dimensions seemed superfluous to physicists. As a matter of fact, the scalar dimension of Hamilton's quaternions was largely ignored. Physicists, especially the famous Maxwell, did pickup on Hamilton's vectors but found no use for the scalar of the quaternion. Hamilton also developed a vector Change function called DEL.

DEL = ∇ = Id/dx + Jd/dy + Kd/dz. applied to any quaternion, e.g. P= Ps + Pv gives:
∇ (Ps +Pv) = - ∇ .Pv + (∇ Ps + ∇ xPv), where ∇ .Pv is called the divergence of Pv and is a scalar; ∇ Ps is called the gradient of Ps and is a vector, and ∇ xPv is called the curl of Pv and is a vector.

Change, the first derivative

With this preliminary introduction we are ready to develop the Living Universe. First the universe is four dimensional and CHANGE in the universe is described by a quaternion Change

2

Operator, symbolized by X for the Ancient Egyptian god of Change, Xepera.

[1] $X = d/cdt + \nabla = d/cdt + Id/dx + Jd/dy + Kd/dz$.

X consists of a scalar part $T = d/cdt$ and a vector part $\nabla = Id/dx + Jd/dy + Kd/dz$.
Hamilton, in spite of the fundamental work he did in mathematics, is probably most famous for his Action function in physics. Hamilton, showed that many things in physics could be simply explained by the changes or lack of changes to a physical concept he called "Action, the product of energy and time", denoted by the symbol "h". Planck's constant was originally known as Planck's constant of action.

In my theory of the Living Universe, Life the product of energy and space is the fundamental concept. Life L is a quaternion and related to action by L=ch:

[2] $L = c (hs + hv) = Ls + ILx + JLy + KLz = Ls + Lv$

Here c is again the speed of light, a scalar. Scalars are mathematically and physically different from vectors. They are commutative and omni-directional. The scalar coefficients (Ls, Lx, Ly and Lz) are in general complex numbers that is have real and imaginary parts. In such cases L was called by Hamilton a Bi-quaternion. Today we would call L a complex quaternion. Complex numbers are a subset of quaternions. Life has units of work-feet, and action has unit's work-seconds. The difference is a question of worldview: is "time" or "space" the dominant dimension in Spacetime? Action says "time" is the worldview. Life says "space" is the proper worldview. The speed of light, transforms time-view variables to space-view variables for example:

$c = L/h = E/B$ giving $L = ch$ and $E = cB$.

E is the space view electric field and B is the time view magnetic field. The E-field and B-field are related by the scalar "c". Multiplying B by the scalar "c" changes only the units not the direction of B. E has the same direction as B, and is identical except for the "scalar units". Maxwell suspected that E and B were related and he called them electromagnetic fields. Maxwell even discovered the relationship between them, "c". He found that the electrostatic units were related to the electromagnetic units by a "constant". When he realized that this "constant" had the same size and units as "c the speed of light", he predicted that electromagnetism and light were the same! This prediction was confirmed by Hertz 20 years later. So paying attention to units pays off.

Change, Work and Quantum Theory
With this conceptual introduction, we can move on to the implications for simplifying mathematics and unifying Relativity and Quantum Physics Theories, until now separate. Energy is one of the fundamental concepts in Physics. Energy also shows the pre quaternion state of physics, where there is no concept of vector energy! Energy is defined as a scalar. The vector product of force and displacement for example is not named vector energy but "torque". Dimensionally, torque has the same units as energy, but the conceptual framework has it

different from energy. In this quaternion framework, torque is directional vector energy as opposed to omni-directional scalar energy. Here energy or work is the first derivative or Change of Life. Mathematically, the Change of Life is Work:

[3] $W = XL = Ws + Wv = (dLs/cdt - \nabla .Lv) + (dLv/cdt + \nabla Ls + \nabla xLv)$

Scalar Work
[3.1] $Ws = (dLs/cdt - \nabla .Lv) = (dhs/dt - \nabla .Lv)$

The scalar Ws is the sum of the radiant of the scalar Ls and the divergence of Lv.
This is isomorphic to Einstein's Photoelectric Equation where: Planck's hsf= dhs/dt and Einstein's "work function, $\Phi = \nabla .Lv$, the divergence of the vector of Life, Lv. It was Einstein's Photoelectric Equation that gave credence to Planck's Quantum Theory.

Planck's Quantum Theory
Planck's Quantum Theory can be derived from the scalar "Continuity Condition":

[3.1.1] $0 = Ws = (dLs/cdt - \nabla .Lv) = (dhs/dt - \nabla .Lv)$

This Scalar Continuity Condition leads to!

[3.1.2] $dhs/dt = \nabla .Lv$

This is Planck's Quantum Theory if dhs/dt equals Planck's "hsf" and $\nabla .Lv$ is the Quantum Energy Φ!

Vector Work
Neither Planck nor Einstein has a vector work equation in their Quantum Theories.
The vector work here derived is Wv:
[3.2] $Wv = (dLv/cdt + \nabla Ls + \nabla xLv)$
The vector work is the sum of the time the radiant of the vector Lv; the gradient of the scalar Ls; and the curl of the vector Lv. These three terms are all vectors and their sum is a vector.
For the Continuity Condition $0 = XL$, the vector equation is:

[3.2.1] $0 = (dLv/cdt + \nabla Ls + \nabla xLv)$

This equation is the vector Continuity Equation and denotes that the three vector terms are dependent and coplanar. Vector Continuity is the basis of Newton's "Action -Reaction Law", Lenz's Law of Electromagnetism and similar "Inertia" laws.
In summary, the first derivative of Life is work W.

Electromagnetic Theory
The Continuity Condition of E, the electric field gives the equations of electromagnetism.

4

[4] $0 = XE = (dEs/cdt - \nabla .Ev) + (dEv/cdt + \nabla Es + \nabla xEv)$

This is the Continuity Equation for the Electric Field E. The Continuity gives two of Maxwell's Equations where $B=E/c$ and $0= \nabla Es$:

[4.1] Coulomb's Law:

Scalar Continuity: $0= (dhs/dt - \nabla .Ev) = (\rho /\varepsilon - \nabla .Ev)$

[4.2] Faraday's Law:

Vector Continuity: $0 = (dEv/cdt + \nabla Es + \nabla xEv) = (dBv/dt + \nabla Es + \nabla xEv)$

Maxwell's other Equations can be derived from $0 =XH$. or by substituting $E=cB=zH=D/e$ in the above. The results give:

[5] $0 = XH = (dHs/cdt - \nabla .Hv) + (dHv/cdt + \nabla Hs + \nabla xHv)$

[5.1] $0 = XH = (dDs/dt - \nabla .Hv) + (dDv/dt + \nabla Hs + \nabla xHv)$

The sign of ∇xHv is opposite from Maxwell's Equations and I believe is the correct sign.

Magnetic Monopole
If we use the B field the existence of the magnetic monopole is shown to be nothing less than the charge density times the impedance z.

$0 = XB = (dBs/cdt - \nabla .Bv) + (dBv/cdt + \nabla Bs + \nabla xBv)$

Where $dBs/cdt = \rho /c\varepsilon = z\rho = \nabla .Bv$. The Magnetic Monopole exists when charge density exists.

Force, Curvature and Relativity Theory
The second derivative of Life has units of Force F. This brings us to Einstein's General Relativity Theory. Again the simplicity of using the Life variable gives the important concepts and equations of physics simply, elegantly and I believe correctly! The second derivative is denoted by X^2 and is the Curvature of Spacetime. Also for simplicity $T=d/cdt$. :

[6] $X^2 = (T^2 - \nabla^2) +2T\nabla$

Note that Curvature has a hyperbolic scalar part.
With these notations Force F is given by:

[7] $F= X^2 L = ((T^2 - \nabla^2) + 2T\nabla) (Ls + Lv) = Fs + Fv$

The scalar force is a scalar/longitudinal wave equation manifesting "gravitational/matter" force.

[7.1] $Fs = ((T^2 - \nabla^2)\, Ls - 2T\nabla\, .Lv)$

The "Cosmological Constant" of Einstein's Relativity Theory naturally appears here as $-2T\nabla$ .Lv, the radiant of the divergence!

[7.2] $Fv = ((T^2 - \nabla^2)\, Lv + 2T(\nabla\, Ls + \nabla\, xLv))$

The vector force Fv is a vector transverse wave, manifesting the electromagnetic effects in the universe. Notice that "matter" is coupled into transverse waves thru Ls, the gradient.
There are two waves in the quaternion equation and they provide an explanation of the Wave Particle Duality!

Equilibrium Condition
The Equilibrium/inflection Equation is given by

[8] $0 = Force = X^2\, L$

Forces at Continuity
At the Continuity Condition $0 = XL$, the Curvature Equation becomes:

[9] $F = -(T^2 + \nabla^2)\, L$

The hyperbolic waves become elliptical at the boundary and represent the extreme values.
If these values are finite the Universe is finite, if infinite, the Universe is infinite. A negative value is the Maximum and a positive value is the Minimum.

Conclusion
This concludes the Living Universe Theory. The variable Life is central to Quantum and Relativity Theory and shows that the two Theories are the first and second derivative of the Life. This theory also shows the universe to be quaternionic in both the variable Life and the CHANGE and CURVATURE functions.

The cosmological Constant is enlivening Astronomy with recent discoveries. I propose that Maxwell's Ether will see a revival in Spacetime. From, my Theory of Life, "free space" is not empty, but is a manifestation of the Ether. The Ether, which consists of two constants:

[10] M= quantum magnetism in voltseconds and

[11] q = quantum charge in Coulombs

These two constant can be determined from two familiar constants: Planck's Constant h = Mq and "free space" impedance zo = M/q. Solving these two equations for the two unknowns give:

[12] M= 500 atto voltseconds and q= 1.326 atto coulombs.

6

The Ancient Egyptians believed that Life was eternal and that we all came from the stars and would return to the stars. We and the stars are part of the Living Universe!

Bibliography:
William Rowan Hamilton
Elements of Quaternions, Chelsea Publishing Company, New York, NY, 1969
James Clerk Maxwell
A treatise on Electricity & Magnetism, Dover Publications, Inc, New York, NY, 1954
Albert Einstein
The Principle of Relativity, Dover Publications, Inc, New York, NY, 1923

UNIFICATION

*Relativity, Electromagnetism and Quantum Theory are united by three constants,
c. Z and h.*

Relativity

Constant 1: The speed of light **c** is the limit speed in Spacetime
This is the essence of Relativity Theory. Einstein said the consequence is of the invariance of c is that spacetime is four-dimensional. I maintain that Spacetime is a four dimensional quaternion space as defined by William Rowan Hamilton.
Hamilton's Vector Rules are: $I^2 = J^2 = K^2 = IJK = -1$
Any displacement in space requires time and the speed is limited by c.
There can be no change in space without a change in time and the speed of the change cannot exceed c, the speed of light.

Units
Light travels 1 foot in a nano second or a speed of 1 Gigafeet/second. The Mile is 6000 feet and the speed of light is 600 million Miles/hour or 10 million Miles/minute.
Let r be a point in Spacetime: $r = ct + Ix + Jy + Kz = R(\cos Q + d\sin Q)$
The Surface $r^2 = (ct)^2 - (x^2 + y^2 + z^2) + 2ct(Ix + Jy + Kz) = R^2(\cos 2Q + d\sin 2Q)$
The scalar is positive when the time segment $(ct)^2 => (x^2 + y^2 + z^2)$.
This will be the case for $c^2 => (x^2 + y^2 + z^2)/t^2 = v^2$ or angle Q less than 45 degrees.
The surface of r^2 degrades to a vector when $\cos 2Q = 0$ or angle Q is 45 degrees
meaning $c^2 = (x^2 + y^2 + z^2)/t^2 = v^2$

Change is the Fundamental Law of the Cosmos.
The Change Operator: $X = d/cdt + \nabla = d/cdt + Id/dx + Jd/dy + Kd/dz = T + \nabla$
The Curvature Operator: $X^2 = (T^2 - \nabla^2) + 2T\nabla$
Change and Curvature are both quaternions as are all points and variables in Spacetime.
Complex numbers are a subset of quaternions.

Electricity

Constant 2: Z is the "free space" impedance of the Ether of the Cosmos.
$Z = m/q$ where **m** is the magnetic charge in webers and **q** is the electric charge in coulombs in free space. (1weber = 1 volt second.)
Z and c relate the four fields of interest in Electricity, the Electric Field E, the Magnetic Density B, the Magnetic Intensity H and the Electric Displacement D .
$E = cB = ZH = cZD$

Electromagnetic fields are quaternions.
$E = Es + IEx + JEy + KEz = Es + Ev$ with a scalar part Es and a vector part Ev.

The Change Equations of Electricity
The Equations governing Electromagnetism are given by the Boundary Condition:
$0 = XB = (dBs/cdt - \nabla.Bv) + (dBv/cdt + \nabla Bs + \nabla xBv)$
Multiplying thru by c gives the Electric Change Equations:

Page 2

$0 = XE = (dBs/dt - \nabla.Ev) + (dBv/dt + \nabla Es + \nabla xEv)$
For a quaternion to be zero, both the scalar and the vector part is zero:
$0 = (dBs/dt - \nabla.Ev)$ **and**
$0 = (dBv/dt + \nabla Es + \nabla xEv)$

The two Magnetic Change Equations are:
$0 = (dDs/dt - \nabla.Hv)$ **and**
$0 = (dDv/dt + \nabla Hs + \nabla xHv)$
Maxwell's Equations are different from these, for example ∇Es, the back emf associated with Lenz's Law is missing from the Electric vector equation. The magnetic vector equation has all the terms but has a negative sign on ∇xHv. Finally, the divergence of the magnetic B-field is not categorically zero, $0 = (dBs/cdt - \nabla.Bv)$.

The Wave Equations:
$X^2 E = ((T^2 - \nabla^2) Es - 2 \nabla.dEv/cdt) + ((T^2 - \nabla^2) Ev + 2d/cdt(\nabla Es + \nabla xEv))$
The Wave Equation at the Boundary or Dependency Condition $(0 = XE)$ is
$X^2 E = - (T^2 + \nabla^2) E$
This wave equation is more complex than Maxwell's , $(T^2 - \nabla^2) Ev$, there are more terms and in fact there are two waves, a scalar wave and a vector wave. The scalar wave manifests "particle-like" behavior and the vector wave , wave-like behavior. The Wave-Particle Duality is the result of two waves. Einstein's Cosmological Constant is seen to be $2 \nabla.dEv/cdt$, twice the divergence of the tangent vector.
$0 = X^2 E$ **defines the Equilibrium Condition internal or at the Boundary.**
The Equilibrium Condition at the Boundary is $0 = (T^2 + \nabla^2) E$.

Quantum Theory

Constant 3: Planck's constant $h = mq$ is the product of the free space charges m and q. Having the value of h and Z allows us to determine the values of the Ethers charges:
662.5 atto atto joule seconds $= mq$ and $120 \pi = 377 = m/q$ giving:
m= 500 atto webers and q= 1.326 atto coulombs (or 8.28 electrons)
The unit of charge in the Ether seems to be 8.28 electrons, while in "matter" the unit of charge is the 1 electron. This indicates a possible difference in the structure of "matter" and the Ether. The Fine Structure Constant $\alpha = 1/137$ may be related to this charge difference. In examining the Hydrogen atoms Quantum speed, $\frac{1}{2}(e/q)^2 = 1/137.12$ appears and is approximately equal to α .

Quantum Theory results from the condition of central forces. Whenever the force F, is in the same direction as the position r, $f \times 2\pi r = 0$ and the force is central. The force being dp/dt makes $\lambda xdp/dt = 0$ or $\lambda xmdv/dt = 0$, or: $2\pi rxp = nh$. **The integer n reflects the boundary conditions, n is an integer for symmetric conditions and (n-½) reflects asymmetric container boundary conditions. The value of n scales to the size of $2\pi rxp/h$.**

Force $= eE = eZcD = eZc(e/4\pi r^2) = mv^2/r = nhv/2\pi r^2$

Velocity $v = e^2 Zc/2hn = c (e/q)^2 /2n = \alpha c/n$
Radius $= n^2 h/2\pi\ m\alpha c$
$En = m(\alpha c/n)^2/2 = 13.6$ eV/n^2

The Photoelectric Equation
The Photoelectric Equation is a Change Equation for a fundamental variable I call Life L. Life L is a quaternion and has the units of energy-feet. Life I related to the variable Action h by c the speed of light L=ch. **The Quantum of Life is hc = 4141 nano eV feet.**

Change of Life is the basis of Quantum Theory:
Work = XL = (dLs/cdt - ∇.Lv) + (dLv/cdt + ∇Ls + ∇xLv)
Substituting h=L/c gives Einstein's Photoelectric Equation:
Work = XL = (dhs/dt - ∇.Lv) + (dhv/dt + ∇Ls + ∇xLv)

The Boundary Condition gives Planck's Quantum Theory:
0 = XL = (dhs/dt - ∇.Lv) + (dhv/dt + ∇Ls + ∇xLv)

This indicates that the "work function" or binding energy is the divergence :
∇.Lv = En = m(αc/n)²/2 = 13.6 eV/n² for Hydrogen atom
Lv= 4.545 eV (Ix + Jy + Kz)/n² and Ls = 4.141 ueVf

The Photon is the scalar/particle energy of the Quantum and there is a vector energy equivalent of Lenz's law, 0 = (dhv/dt + ∇Ls + ∇xLv).

Relativity Theory

Force = X² L is the second derivative or Curvature of L:
X² L = ((T² - ∇ ²) Ls - 2 ∇.dLv/cdt) + ((T² - ∇ ²) Lv + 2d/cdt (∇Ls + ∇xLv))
The Wave -Particle Duality arises again as the scalar wave reflects "Particle" longitudinal (2 ∇ .dLv/cdt) aspects and the vector wave reflects transverse (∇xLv) "wave" aspects. Gravity and the Electric Field are reflected in this single quaternion wave equation. Einstein's Cosmological Constant is seen in the longitudinal (2 ∇.dLv/cdt) part of the scalar wave.

I believe the Cosmos is bounded, 0= XL and thus at its boundary the force is:
Force = - X² L= - (T² + ∇ ²) L

Summary
The constants c, Z and h have been shown to be key elements in understanding the Cosmos. Electricity, Relativity and Quantum Theory are bound together and Unified by these constants. The implications of this unification are still to be developed, but the trend is toward simplification and understanding.

On charged bodies in constant magnetic fields
© 2004 by Wardell Lindsay

On charged bodies in constant magnetic fields © 2004 by wardell lindsay

A charged body moving in a constant magnetic field has energy = ½ mv^2. The motion of the body depends on the force from the field. The force will change the velocity of the body by, v= f/m t. The force on the moving charge is f = qvB = qvBcosa + qvBsina, where "a" is the angle between the velocity and the B-field. If the angle is zero, there is no circulation of the charge around the B-filed, only oscillation along the B-field. If the angle is 90 degrees, there is no oscillation along the B-field, only circulation around the B-field.

Let's examine the force and motion changes, for example let the angle be 90 degrees and a positive charge + moving in the "I" direction and in a constant B-field in the "J" direction.

F = qvB
+K = +I J
-I = +KJ
-K= -IJ
+I = -KJ
+K = +IJ
Etc

This example shows the circular motion of the charge in a constant field. Plus charges circulate clockwise (cw) and negative charges circulate counter cw.

For a charge moving with angle zero degrees v=J, then the force is not circulation but "spring-like" oscillation. **The magnetic field acts like a spring for parallel charge motion!**

F = qvB
-1 = +JJ
-J = +-1J
+1 = +-JJ
+J = ++1J
-1 = ++JJ
Etc

Here we see the motion is confined to oscillation along the J direction.

If the angle is not zero or 90 degrees, the charge circulates and oscillates or orbits in a plane containing the original velocity vector. For example if v= I + J then the orbit contains the points (I + J), (-1 + K), - (I + J) and (1 -K). This motion could explain "magnetic trapping". The charge is confined to an orbit plane in the constant B-field.

The energy of the charge is seen to be conserved: ½ mv = ½ mv²cosa² + ½mv² sina², where vcosa is the parallel velocity and vsina is the transverse perpendicular velocity.

Here is a drawing of the orbit produced by the magnetic field and moving charge.

2

The orbit of the gyration does not lie in the j-plane but intersects the j-plane and the charge oscillates in the direction of the j-plane as it orbit's the j-axis. The **vibration along the j-axis** is shown in the (complex) Scalar Product plane and the **gyration around the j-axis** is shown in the Vector product plane.

www.ingramcontent.com/pod-product-compliance
Lightning Source LLC
Chambersburg PA
CBHW021931170526
45157CB00005B/2279